Fractions Bingo Book

COMPLETE BINGO GAME IN A BOOK

Written By Rebecca Stark

Educational Books 'n' Bingo

TITLE: Fractions Bingo
AUTHOR: Rebecca Stark

ISBN 978-0-87386-431-2

Educational Books 'n' Bingo

Printed in the U.S.A.

FRACTIONS BINGO DIRECTIONS

INCLUDED:

List of Terms

Templates for Additional Terms and Clues

2 Clues per Term

30 Unique Bingo Cards

Markers

1. **Either cut apart the book or make copies of ALL the sheets. You might want to make an extra copy of the clue sheets to use for introduction and review. Keep the sheets in an envelope for easy reuse.**

2. Cut apart the call cards with terms and clues.

3. Pass out one bingo card per student. There are enough for a class of 30.

4. Pass out markers. You may cut apart the markers included in this book or use any other small items of your choice.

5. Decide whether or not you will require the entire card to be filled. Requiring the entire card to be filled provides a better review. However, if you have a short time to fill, you may prefer to have them do the just the border or some other format. Tell the class before you begin what is required.

6. There are 50 terms. Read the list before you begin. If there are any terms that have not been covered in class, you may want to read to the students the term and clues before you begin.

7. There is a blank space in the middle of each card. You can instruct the students to use it as a free space or you can write in answers to cover terms not included. Of course, in this case you would create your own clues. (Templates provided.)

8. Shuffle the cards and place them in a pile. Two or three clues are provided for each term. If you plan to play the game with the same group more than once, you might want to choose a different clue for each game. If not, you may choose to use more than one clue.

9. Be sure to keep the cards you have used for the present game in a separate pile. When a student calls, "Bingo," he or she will have to verify that the correct answers are on his or her card AND that the markers were placed in response to the proper questions. Pull out the cards that are on the student's card keeping them in the order they were used in the game. Read each clue as it was given and ask the student to identify the correct answer from his or her card.

10. If the student has the correct answers on the card AND has shown that they were marked in response to the *correct questions,* then that student is the winner and the game is over. If the student does not have the correct answers on the card OR he or she marked the answers in response to *the wrong questions,* then the game continues until there is a proper winner.

11. If you want to play again, reshuffle the cards and begin again.

Have fun!

TERMS/ANSWERS INCLUDED

1/5	4/5
25%	5/6
2/9	6/7
3/10	7/8
1/3	8/9
3/8	11/9
4/10	9/4
2/5	12/5
5/12	2 3/4
3/7	15/4
4/9	4 1/3
50%	6 1/2
.50	COMMON DENOMINATOR
1/2	DECIMAL
7/13	DENOMINATOR
5/9	EQUIVALENT
4/7	FRACTION
60%	IMPROPER FRACTION
3/5	MIXED NUMBER
5/8	NUMERATOR
2/3	PROPER FRACTION
5/7	RECIPROCAL
8/11	REDUCE
3/4	SIMPLEST FORM
7/9	%

Additional Terms

Choose as many fractions terms as you would like and write them in the squares. Repeat each as desired.
Cut out the squares and randomly distribute them to the class.
Instruct the students to place their square on the center space of their card.

Clues for Additional Terms

Write three clues for each of your additional terms.

_____	_____
1.	1.
2.	2.
3.	3.
_____	_____
1.	1.
2.	2.
3.	3.
_____	_____
1.	1.
2.	2.
3.	3.

Fractions Bingo

1/2	1/2	1/2	1/2	1/2
1/2	1/2	1/2	1/2	1/2
1/2	1/2	1/2	1/2	1/2
1/2	1/2	1/2	1/2	1/2
1/2	1/2	1/2	1/2	1/2
1/2	1/2	1/2	1/2	1/2
1/2	1/2	1/2	1/2	1/2

1/5 1. This fraction is read "one-fifth." 2. 2/5 − 1/5 = ___ 3. An equivalent fraction is 2/10.	**25%** 1. This is read "twenty-five percent." 2. Written as a fraction, this percentage is 25/100, which is equal to 1/4. 3. If this percentage were written as a decimal, it would be .25.
2/9 1. This fraction is read "two-ninths." 2. 1/9 + 1/9 = ___ 3. 3/9 − 1/9 = ___	**3/10** 1. This fraction is read "three-tenths." 2. 1/10 + 2/10 = ___ 3. 4/10 − 1/10 = ___
1/3 1. This fraction is read "one-third." 2. 2/3 − 1/3 = ___ 3. If you divide a pie into 3 equal parts, one of those parts would be ___ of the pie.	**3/8** 1. This fraction is read "three-eighths." 2. 2/8 + 1/8 = ___ 3. 5/8 − 2/8 = ___
4/10 1. This fraction is read "four-tenths." 2. 2/10 + 2/10 = ___ 3. An equivalent fraction is 2/5.	**2/5** 1. This fraction is read "two-fifths." 2. 1/5 + 1/5 = ___ 3. 4/5 − 2/5 = ___
5/12 1. This fraction is read "five-twelfths." 2. 2/12 + 3/12 = ___ 3. 7/12 − 2/12 = ___	**3/7** 1. This fraction is read "three-sevenths." 2. 2/7 + 1/7 = ___ 3. 6/7 − 3/7 = ___

Fractions Bingo

4/9 1. This fraction is read "four-ninths." 2. 2/9 + 2/9 = ___ 3. 8/9 − 4/9 = ___	**50%** 1. This is read "fifty percent." 2. Written as a fraction, this percentage is 50/100, which is equal to 1/2. 3. If this percentage were written as a decimal, it would be .50.
.50 1. This decimal number is read as "fifty hundredths." 2. Written as a percentage, this decimal number would be 50%. 3. We could also write this decimal number as .5, or 5 tenths.	**1/2** 1. This fraction is read "one-half." 2. 1/4 + 1/4 = 2/4 or ___ 3. If you divide a pie into 2 equal parts, one of those parts would be ___ of the pie.
7/13 1. This fraction is read "seven-thirteenths." 2. 2/13 + 5/13 = ___ 3. 11/13 − 4/13 = ___	**5/9** 1. This fraction is read "five-ninths." 2. 4/9 + 1/9 = ___ 3. 7/9 − 2/9 = ___
4/7 1. This fraction is read "four-sevenths." 2. 2/7 + 2/7 = ___ 3. 5/7 − 1/7 = ___	**60%** 1. This is read "sixty percent." 2. Written as a fraction, this percentage is 60/100, which is equal to 3/5. 3. If this percentage were written as a decimal, it would be .60.
3/5 1. This fraction is read "three-fifths." 2. 2/5 + 1/5 = ___ 3. An equivalent fraction is 6/10.	**5/8** 1. This fraction is read "five-eighths." 2. 3/8 + 2/8 = ___ 3. 7/8 − 2/8 = ___

Fractions Bingo

2/3 1. This fraction is read "two-thirds." 2. $1/3 + 1/3 =$ ___ 3. An equivalent fraction is 4/6.	**5/7** 1. This fraction is read "five-sevenths." 2. $2/7 + 3/7 =$ ___ 3. $6/7 - 1/7 =$ ___
8/11 1. This fraction is read "eight-elevenths." 2. $2/11 + 6/11 =$ ___ 3. $10/11 - 2/11 =$ ___	**3/4** 1. This fraction is read "three-fourths." 2. $2/4 + 1/4 =$ ___ 3. ___ $- 2/4 = 1/4$
7/9 1. This fraction is read "seven-ninths." 2. $4/9 + 3/9 =$ ___ 3. $8/9 - 1/9 =$ ___	**4/5** 1. This fraction is read "four-fifths." 2. $3/5 + 1/5 =$ ___ 3. An equivalent fraction is 8/10.
5/6 1. This fraction is read "five-sixths." 2. $4/6 + 1/6 =$ ___ 3. ___ $- 1/6 = 4/6$	**6/7** 1. This fraction is read "six-sevenths." 2. $4/7 + 2/7 =$ ___ 3. ___ $- 1/7 = 5/7$
7/8 1. This fraction is read "seven-eighths." 2. $4/8 + 3/8 =$ ___ 3. ___ $- 2/8 = 5/8$	**8/9** 1. This fraction is read "eight-ninths." 2. $4/9 + 4/9 =$ ___ 3. ___ $- 2/9 = 6/9$

Fractions Bingo

11/9 1. This improper fraction is read as "eleven-ninths." 2. To write this improper fraction as a mixed number, you would write 1 2/9. 3. 3/9 + 8/9 = ___, which is an improper fraction.	**9/4** 1. This improper fraction is read as "nine-fourths." 2. To write this improper fraction as a mixed number, you would write 1 2/9. 3. 3/9 + 8/9 = ___, which is an improper fraction.
12/5 1. This improper fraction is read as "twelve-fifths." 2. To write this improper fraction as a mixed number, you would write 2 2/5. 3. 3/9 + 8/9 = ___, which is an improper fraction.	**2 3/4** 1. This mixed number is read as "two and three-fourths." 2. To write this mixed number as an improper fraction, you would write 11/4. 3. 2 + 3/4 = ___, which is a mixed
15/4 1. This improper fraction is read as "fifteen-fourths." 2. To write this improper fraction as a mixed number, you would write 3 3/4. 3. 1/4 + 14/4 = ___, which is an improper fraction.	**4 1/3** 1. This mixed number is read as "four and one-third." 2. To write this mixed number as an improper fraction, you would write 13/3. 3. 4 + 1/3 = ___, which is a mixed
6 1/2 1. This mixed number is read as "six and one-half." 2. To write this mixed number as an improper fraction, you would write 13/2. 3. 6 1/4 + 1/4 = 6 2/4 or ___.	**Common Denominator** 1. If two or more fractions have the same denominator, we say they have a ___. 2. The least, or lowest, ___ of 1/2 and 1/4 is 4. 3. The least, or lowest, ___ of 1/2
Decimal 1. A ___ number is one having one or more places to the right of a decimal point. 2. 36.85 is one. 3. 87.2 is one.	**Denominator** 1. It is what we call the bottom number of a fraction. 2. In the fraction 5/7, 7 is the ___. 3. In the fraction 9/11, 11 is the ___.

Equivalent 1. ___ fractions have the same value. 2. 1/2 and 2/4 are ___ fractions. 3. 2/3, 4/6 and 8/12 are ___ fractions.	**Fraction** 1. It is a number that expresses part of a group. 2. 2/3 is a proper one; 3/2 is an improper one. 3. Two parts of a ___ are the numerator and the denominator.
Improper Fraction 1. An ___ is one whose numerator is larger than its denominator. 2. 4/3 is one, but 3/4 is not. 3. 22/13 is one, but 13/22 is not.	**Mixed Number** 1. This kind of number has a whole number part and a fractional part. 2. 5 4/7 is one, but 4/7 is not. 3. 9 2/5 is one, but 2/5 is not.
Numerator 1. It is what we call the top number of a fraction. 2. In the fraction 5/7, 5 is the ___. 3. In the fraction 9/11, 9 is the ___.	**Proper Fraction** 1. A ___ is one whose denominator is larger than its numerator. 2. 4/9 is one, but 9/4 is not. 3. 2/5 is one, but 5/2 is not.
Reciprocal 1. 2/3 is the ___ of 3/2. 2. 4/5 is the ___ of 5/4. 3. 7 is the ___ of 1/7.	**Reduce** 1. To ___ is to simplify the form of a fraction without changing its value. 2. If we ___ the fraction 2/4, we get 1/2. 3. If we ___ the fraction 4/6, we get 2/3.
Simplest Form 1. When we reduce a fraction as much as possible, we say it is in its ___. 2. 8/16, 5/10 and 10/20 are equivalent fractions. Their ___ is 1/2. 3. 2/6, 4/12 and 5/15 are equivalent	**%** 1. This is the symbol for percent. 2. If 1/2 the students in your class are boys, then 50___ of the students are boys. 3. If 1/4 of the students in your class are girls, then 25___ of the students

Fractions Bingo

Fractions Bingo

$\dfrac{4}{5}$	$\dfrac{5}{7}$	Fraction	Simplest Form	50%
$\dfrac{5}{12}$	$\dfrac{1}{5}$	Reduce	$\dfrac{9}{4}$	$\dfrac{7}{9}$
Equivalent	$6\dfrac{1}{2}$		$\dfrac{6}{7}$	$\dfrac{12}{5}$
%	Common Denominator	$\dfrac{5}{8}$	Proper Fraction	$\dfrac{5}{6}$
$2\dfrac{3}{4}$	$\dfrac{1}{2}$	$\dfrac{3}{7}$	60%	$\dfrac{2}{3}$

Fractions Bingo: Card No. 1

Fractions Bingo

%	Denominator	$\dfrac{7}{8}$	$\dfrac{15}{4}$	$2\dfrac{3}{4}$
$\dfrac{5}{6}$	$\dfrac{9}{4}$	$\dfrac{1}{3}$	Common Denominator	Improper Fraction
Decimal	$\dfrac{1}{2}$		$\dfrac{4}{9}$	$\dfrac{5}{8}$
$\dfrac{4}{7}$	25%	$6\dfrac{1}{2}$	$\dfrac{3}{5}$	$\dfrac{7}{9}$
$\dfrac{2}{3}$	Reduce	$\dfrac{3}{7}$	$\dfrac{5}{12}$	60%

Fractions Bingo: Card No. 2

Fractions Bingo

%	$\dfrac{5}{8}$	$\dfrac{9}{4}$	Proper Fraction	Equivalent
$\dfrac{1}{2}$	$\dfrac{1}{5}$	$\dfrac{3}{8}$	$\dfrac{5}{7}$	$\dfrac{3}{4}$
Common Denominator	Reduce		Improper Fraction	$\dfrac{2}{9}$
$6\dfrac{1}{2}$	Decimal	$2\dfrac{3}{4}$	$\dfrac{4}{7}$	$\dfrac{7}{8}$
60%	$\dfrac{5}{12}$	$\dfrac{3}{7}$	$\dfrac{3}{5}$	Fraction

Fractions Bingo

$6\frac{1}{2}$	Improper Fraction	$2\frac{3}{4}$	$\frac{5}{12}$	Mixed Number
$\frac{8}{9}$	$\frac{3}{10}$	$\frac{5}{7}$	$\frac{15}{4}$	Equivalent
$\frac{6}{7}$	$\frac{4}{7}$		50%	Simplest Form
$\frac{5}{8}$	Denominator	Reduce	$\frac{3}{7}$	$\frac{3}{8}$
$\frac{8}{11}$	$\frac{2}{3}$	$\frac{11}{9}$	60%	$\frac{12}{5}$

Fractions Bingo: Card No. 4

Fractions Bingo

$\dfrac{2}{3}$	50%	**Common Denominator**	$\dfrac{1}{3}$	$\dfrac{5}{12}$
$\dfrac{8}{9}$	$\dfrac{5}{8}$	$\dfrac{3}{8}$	$\dfrac{4}{9}$	$\dfrac{1}{5}$
Equivalent	$\dfrac{12}{5}$		.50	$\dfrac{7}{13}$
$\dfrac{7}{9}$	**Improper Fraction**	$\dfrac{4}{5}$	$\dfrac{3}{5}$	$\dfrac{8}{11}$
$\dfrac{9}{4}$	$\dfrac{3}{7}$	**Denominator**	$6\dfrac{1}{2}$	$\dfrac{6}{7}$

Fractions Bingo

$\dfrac{2}{9}$	**Improper Fraction**	$\dfrac{7}{8}$	**Fraction**	$\dfrac{12}{5}$
Proper Fraction	**Common Denominator**	$\dfrac{8}{11}$	$\dfrac{5}{7}$	**Equivalent**
$\dfrac{15}{4}$	$\dfrac{4}{10}$		$\dfrac{3}{10}$	$\dfrac{4}{9}$
$\dfrac{3}{7}$	$2\dfrac{3}{4}$	$\dfrac{3}{5}$	$\dfrac{11}{9}$	**Mixed Number**
$\dfrac{5}{6}$	$\dfrac{5}{8}$	$\dfrac{4}{5}$	$\dfrac{6}{7}$	**Denominator**

Fractions Bingo: Card No. 6

Fractions Bingo

$\dfrac{4}{5}$	**Improper Fraction**	$\dfrac{7}{13}$	.50	$\dfrac{9}{4}$
$\dfrac{5}{6}$	$2\dfrac{3}{4}$	$\dfrac{1}{2}$	$\dfrac{1}{5}$	$\dfrac{8}{9}$
$\dfrac{7}{8}$	**Simplest Form**		$\dfrac{4}{9}$	$\dfrac{3}{10}$
$6\dfrac{1}{2}$	$\dfrac{4}{7}$	$\dfrac{3}{8}$	**%**	**Decimal**
$\dfrac{3}{7}$	$\dfrac{5}{12}$	$\dfrac{3}{5}$	$\dfrac{11}{9}$	$\dfrac{2}{9}$

Fractions Bingo

$\dfrac{6}{7}$	Improper Fraction	$\dfrac{2}{5}$	Proper Fraction	$\dfrac{3}{10}$
$\dfrac{8}{9}$	Fraction	$\dfrac{15}{4}$	$\dfrac{12}{5}$	$\dfrac{1}{3}$
Equivalent	$4\dfrac{1}{3}$		Mixed Number	50%
60%	$6\dfrac{1}{2}$	%	$\dfrac{8}{11}$	$\dfrac{4}{7}$
Reduce	$\dfrac{3}{7}$	$\dfrac{11}{9}$	Common Denominator	$\dfrac{5}{6}$

Fractions Bingo

$\dfrac{4}{9}$	$\dfrac{9}{4}$	$\dfrac{1}{2}$	**Equivalent**	$\dfrac{5}{12}$
$\dfrac{8}{11}$	**Fraction**	$\dfrac{6}{7}$	**Common Denominator**	**Mixed Number**
$\dfrac{3}{4}$	$\dfrac{4}{5}$		$\dfrac{1}{5}$	$\dfrac{2}{5}$
$\dfrac{4}{10}$	$\dfrac{2}{3}$	$2\dfrac{3}{4}$	.50	$\dfrac{7}{13}$
$\dfrac{4}{7}$	$\dfrac{3}{5}$	$\dfrac{3}{8}$	%	50%

Fractions Bingo: Card No. 9

Fractions Bingo

%	Proper Fraction	$\dfrac{3}{10}$	$\dfrac{15}{4}$	Equivalent
$\dfrac{12}{5}$	$\dfrac{1}{3}$	$\dfrac{5}{7}$	$\dfrac{1}{5}$	Mixed Number
$4\dfrac{1}{3}$	Improper Fraction		Simplest Form	Decimal
$2\dfrac{3}{4}$	$\dfrac{7}{9}$	$\dfrac{8}{11}$	$\dfrac{3}{5}$	$\dfrac{3}{4}$
$\dfrac{3}{8}$	$\dfrac{5}{6}$	$\dfrac{7}{8}$	$\dfrac{2}{3}$	$\dfrac{6}{7}$

Fractions Bingo: Card No. 10

Fractions Bingo

$\dfrac{2}{9}$	**Improper Fraction**	**Common Denominator**	$\dfrac{8}{11}$	$\dfrac{5}{6}$
$\dfrac{2}{5}$	$\dfrac{3}{4}$	.50	$\dfrac{4}{9}$	$\dfrac{5}{7}$
$\dfrac{8}{9}$	**Fraction**		$\dfrac{7}{8}$	$\dfrac{1}{2}$
$\dfrac{3}{8}$	**Equivalent**	$\dfrac{3}{5}$	$\dfrac{5}{12}$	%
$\dfrac{4}{10}$	$\dfrac{3}{7}$	$\dfrac{4}{5}$	$\dfrac{11}{9}$	$\dfrac{9}{4}$

Fractions Bingo: Card No. 11

Fractions Bingo

$\dfrac{9}{4}$	50%	$\dfrac{3}{4}$	Proper Fraction	$\dfrac{4}{9}$
$\dfrac{1}{2}$	Reduce	Fraction	$\dfrac{11}{9}$	$\dfrac{1}{5}$
$\dfrac{4}{5}$	$\dfrac{7}{13}$		$\dfrac{12}{5}$	$\dfrac{15}{4}$
$\dfrac{3}{7}$	$\dfrac{4}{7}$	Mixed Number	%	$\dfrac{8}{9}$
Improper Fraction	$\dfrac{2}{5}$	$4\dfrac{1}{3}$	$\dfrac{4}{10}$	$\dfrac{1}{3}$

Fractions Bingo

$\dfrac{4}{10}$	50%	$\dfrac{2}{9}$	$\dfrac{3}{4}$	$\dfrac{12}{5}$
Fraction	$\dfrac{2}{5}$	Improper Fraction	$\dfrac{4}{9}$	Decimal
Proper Fraction	$\dfrac{1}{3}$		$\dfrac{1}{2}$	$\dfrac{7}{13}$
$\dfrac{6}{7}$	$\dfrac{3}{5}$	$\dfrac{3}{10}$	$4\dfrac{1}{3}$	%
$\dfrac{3}{7}$	$\dfrac{7}{9}$	$\dfrac{11}{9}$	$\dfrac{4}{5}$	.50

Fractions Bingo

$\dfrac{5}{12}$	**Fraction**	**Common Denominator**	$\dfrac{4}{9}$	$\dfrac{4}{10}$
$\dfrac{1}{3}$	$\dfrac{4}{5}$	$\dfrac{3}{4}$	$\dfrac{1}{5}$	**Improper Fraction**
$\dfrac{8}{11}$	**Simplest Form**		$\dfrac{7}{8}$	$\dfrac{3}{8}$
$\dfrac{7}{9}$	$\dfrac{3}{5}$	$4\dfrac{1}{3}$	$\dfrac{3}{10}$	$\dfrac{2}{9}$
$\dfrac{3}{7}$	$\dfrac{15}{4}$	**Decimal**	$\dfrac{5}{6}$	$\dfrac{6}{7}$

Fractions Bingo: Card No. 14

Fractions Bingo

.50	$\frac{4}{9}$	Common Denominator	$\frac{9}{4}$	Proper Fraction
$\frac{2}{9}$	$\frac{7}{8}$	$\frac{5}{7}$	Fraction	$\frac{8}{11}$
$\frac{12}{5}$	$\frac{4}{5}$		Equivalent	Mixed Number
$\frac{3}{7}$	$\frac{3}{4}$	$\frac{2}{5}$	$\frac{3}{5}$	$\frac{4}{10}$
$\frac{5}{6}$	$\frac{4}{7}$	$\frac{11}{9}$	Denominator	$\frac{1}{2}$

Fractions Bingo: Card No. 15

Fractions Bingo

$\dfrac{3}{10}$	$\dfrac{3}{4}$	$\dfrac{2}{5}$	Denominator	25%
$\dfrac{15}{4}$	Decimal	$\dfrac{7}{13}$	$\dfrac{8}{9}$	Simplest Form
$\dfrac{4}{10}$	50%		$\dfrac{12}{5}$	$\dfrac{1}{2}$
$6\dfrac{1}{2}$	$\dfrac{1}{3}$	$\dfrac{3}{7}$	.50	%
$\dfrac{8}{11}$	Reciprocal	$\dfrac{11}{9}$	$\dfrac{4}{7}$	Improper Fraction

Fractions Bingo: Card No. 16

Fractions Bingo

$\frac{3}{8}$	Numerator	$\frac{5}{9}$	$\frac{3}{4}$	$\frac{5}{12}$
.50	$\frac{8}{11}$	$\frac{3}{5}$	Simplest Form	$\frac{7}{13}$
$\frac{4}{9}$	$\frac{6}{7}$		Reciprocal	$\frac{2}{5}$
$\frac{2}{3}$	$\frac{5}{6}$	%	Common Denominator	Decimal
$2\frac{3}{4}$	$\frac{4}{10}$	$\frac{9}{4}$	Proper Fraction	50%

Fractions Bingo

Denominator	$4\frac{1}{3}$	$\frac{1}{3}$	$\frac{8}{11}$	$\frac{15}{4}$
Improper Fraction	$\frac{3}{8}$	$2\frac{3}{4}$	$\frac{12}{5}$	$\frac{4}{10}$
$\frac{4}{9}$	Decimal		$\frac{5}{9}$	Mixed Number
$\frac{2}{3}$	$\frac{5}{7}$	$\frac{3}{5}$	%	$\frac{7}{8}$
Reciprocal	$\frac{3}{4}$	Common Denominator	Numerator	$\frac{2}{9}$

Fractions Bingo: Card No. 18

Fractions Bingo

$\dfrac{12}{5}$	$\dfrac{2}{9}$	$\dfrac{3}{4}$	$\dfrac{2}{5}$	$4\dfrac{1}{3}$
.50	Proper Fraction	Mixed Number	$\dfrac{9}{4}$	Simplest Form
Numerator	$\dfrac{5}{12}$		$\dfrac{1}{5}$	Denominator
$\dfrac{7}{8}$	Reciprocal	$2\dfrac{3}{4}$	$\dfrac{4}{7}$	$\dfrac{5}{9}$
Equivalent	25%	$\dfrac{5}{6}$	$\dfrac{6}{7}$	$\dfrac{11}{9}$

Fractions Bingo: Card No. 19

Fractions Bingo

$4\frac{1}{3}$	**Numerator**	**Proper Fraction**	$\frac{3}{4}$	$\frac{1}{5}$
$\frac{1}{3}$	$\frac{1}{2}$	$\frac{8}{9}$	$2\frac{3}{4}$	$\frac{15}{4}$
50%	$\frac{7}{13}$		$6\frac{1}{2}$	$\frac{5}{7}$
$\frac{2}{3}$	**Reduce**	**60%**	$\frac{4}{7}$	**Reciprocal**
$\frac{5}{8}$	$\frac{6}{7}$	**25%**	**%**	$\frac{5}{9}$

Fractions Bingo: Card No. 20

Fractions Bingo

.50	$\dfrac{2}{9}$	$\dfrac{8}{9}$	$\dfrac{3}{4}$	$\dfrac{7}{9}$
50%	$\dfrac{5}{9}$	$\dfrac{3}{10}$	$\dfrac{2}{5}$	$\dfrac{4}{5}$
Decimal	$\dfrac{5}{6}$		Numerator	Common Denominator
$2\dfrac{3}{4}$	$\dfrac{9}{4}$	Reciprocal	$\dfrac{2}{3}$	$\dfrac{6}{7}$
$6\dfrac{1}{2}$	25%	$\dfrac{11}{9}$	$\dfrac{3}{8}$	$\dfrac{4}{7}$

Fractions Bingo: Card No. 21

Fractions Bingo

Equivalent	$\dfrac{7}{8}$	$\dfrac{5}{9}$	Fraction	$\dfrac{4}{10}$
$\dfrac{15}{4}$	Proper Fraction	Denominator	$\dfrac{2}{5}$	$\dfrac{1}{5}$
$\dfrac{1}{3}$	Simplest Form		$\dfrac{4}{5}$	$\dfrac{7}{13}$
Reciprocal	$\dfrac{2}{3}$	$\dfrac{4}{7}$	$\dfrac{5}{7}$	$\dfrac{5}{12}$
25%	$\dfrac{3}{8}$	Numerator	Decimal	$\dfrac{8}{9}$

Fractions Bingo: Card No. 22

Fractions Bingo

$\dfrac{3}{10}$	**Numerator**	$\dfrac{9}{4}$	**Fraction**	$\dfrac{11}{9}$
$\dfrac{2}{9}$	$4\dfrac{1}{3}$	$\dfrac{5}{6}$	.50	$\dfrac{5}{7}$
$\dfrac{7}{8}$	$\dfrac{4}{10}$		60%	$\dfrac{4}{5}$
Decimal	25%	**Reciprocal**	$\dfrac{3}{8}$	$\dfrac{4}{7}$
$\dfrac{7}{9}$	**Reduce**	$\dfrac{6}{7}$	$2\dfrac{3}{4}$	$\dfrac{5}{9}$

Fractions Bingo

$\dfrac{3}{10}$	$4\dfrac{1}{3}$	$\dfrac{5}{12}$	Numerator	$\dfrac{2}{5}$
$\dfrac{12}{5}$	$\dfrac{11}{9}$	$\dfrac{8}{9}$	$\dfrac{15}{4}$	$\dfrac{4}{5}$
$\dfrac{7}{13}$	Denominator		$\dfrac{4}{10}$	Decimal
$\dfrac{7}{9}$	60%	Reciprocal	$\dfrac{3}{8}$	50%
$\dfrac{5}{8}$	$6\dfrac{1}{2}$	25%	Proper Fraction	Reduce

Fractions Bingo

$6\frac{1}{2}$	$\frac{8}{9}$	Numerator	Common Denominator	$\frac{5}{9}$
$\frac{5}{7}$	$\frac{7}{9}$	.50	$\frac{3}{10}$	$\frac{1}{5}$
50%	$\frac{2}{5}$		60%	Reciprocal
Denominator	$\frac{2}{3}$	Reduce	25%	Simplest Form
$\frac{11}{9}$	$\frac{5}{12}$	$\frac{1}{3}$	$\frac{8}{11}$	$\frac{5}{8}$

Fractions Bingo

$\frac{5}{9}$	**Numerator**	$\frac{7}{8}$	$\frac{15}{4}$	**Denominator**
$2\frac{3}{4}$	**Proper Fraction**	$\frac{2}{5}$	$4\frac{1}{3}$	$\frac{3}{10}$
$\frac{7}{9}$	**60%**		**Simplest Form**	$6\frac{1}{2}$
$\frac{3}{8}$	**Fraction**	$\frac{2}{3}$	**25%**	**Reciprocal**
$\frac{7}{13}$	$\frac{8}{11}$	**Common Denominator**	**Reduce**	$\frac{5}{8}$

Fractions Bingo

$\dfrac{7}{8}$	$\dfrac{1}{3}$	**Numerator**	$4\dfrac{1}{3}$	$\dfrac{1}{2}$
$\dfrac{7}{9}$	**60%**	**.50**	**Reciprocal**	$\dfrac{1}{5}$
$\dfrac{3}{5}$	**Reduce**		**25%**	$6\dfrac{1}{2}$
Denominator	$\dfrac{2}{9}$	$\dfrac{8}{9}$	$\dfrac{5}{8}$	$\dfrac{5}{7}$
$\dfrac{4}{10}$	**Simplest Form**	$\dfrac{5}{9}$	**Equivalent**	$\dfrac{7}{13}$

Fractions Bingo: Card No. 27

Fractions Bingo

$\dfrac{12}{5}$	$4\dfrac{1}{3}$	**Denominator**	**Numerator**	$\dfrac{3}{10}$
$\dfrac{1}{2}$	$\dfrac{5}{9}$	**60%**	$\dfrac{15}{4}$	**Simplest Form**
Reduce	**Decimal**		$\dfrac{7}{13}$	$2\dfrac{3}{4}$
%	**Equivalent**	$\dfrac{5}{6}$	**25%**	**Reciprocal**
Fraction	$\dfrac{4}{9}$	$\dfrac{4}{10}$	$\dfrac{5}{8}$	$\dfrac{7}{9}$

Fractions Bingo

$\dfrac{5}{9}$	$4\dfrac{1}{3}$	Denominator	.50	$\dfrac{4}{9}$
$\dfrac{7}{9}$	$2\dfrac{3}{4}$	$\dfrac{8}{9}$	$\dfrac{7}{13}$	Equivalent
50%	60%		$\dfrac{1}{5}$	Numerator
$\dfrac{1}{2}$	$\dfrac{2}{3}$	Fraction	25%	Reciprocal
$\dfrac{3}{10}$	$\dfrac{2}{5}$	$\dfrac{5}{8}$	$\dfrac{2}{9}$	Reduce

Fractions Bingo: Card No. 29

Fractions Bingo

$\dfrac{5}{12}$	**Numerator**	$\dfrac{15}{4}$	$\dfrac{4}{9}$	**Reciprocal**
$\dfrac{5}{7}$	$4\dfrac{1}{3}$	$\dfrac{7}{8}$	**Simplest Form**	$\dfrac{1}{5}$
$\dfrac{7}{9}$	$\dfrac{4}{10}$		$\dfrac{7}{13}$	$\dfrac{8}{9}$
$\dfrac{5}{8}$	$\dfrac{2}{9}$	**Fraction**	**25%**	**60%**
$\dfrac{2}{3}$	$\dfrac{9}{4}$	**Reduce**	$\dfrac{5}{9}$	**Denominator**

Fractions Bingo: Card No. 30